Precipitation

by Stephen Schutz

Starfall®

Starfall Education, P.O. Box 359, Boulder, CO 80306

ISBN: 978-1-59577-061-5

Do you like the **rain**?
Some people don't.

But…

…some people do like the rain!

Do you like the **snow**?
Some people don't.

But…

...some people do like the snow!

Hail is different from snow.
You may have seen hail like this.

But…

…have you ever seen hail like this?

Rain, snow, and hail all fall from the sky. They are called **precipitation**.

How is precipitation made?

Precipitation comes from water in the air.

Have you ever seen rain puddles disappear? When puddles dry up, the water goes into the air!

The air is filled with water that we cannot see. We call this **water vapor**.

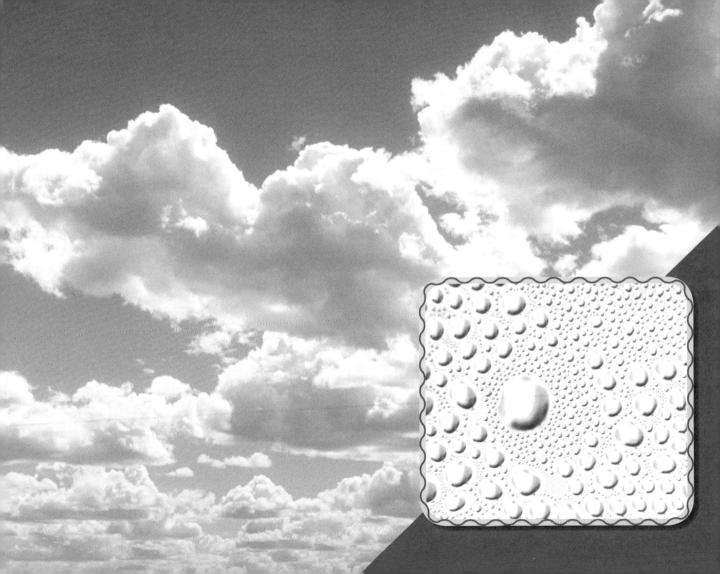

Water vapor in the air can change into water droplets or ice crystals.

This is how **clouds** are made!

Water falls from clouds as precipitation. Precipitation can be rain, snow, or hail.

Which one do *you* like best?

Precipitation

Rain

Snow

Hail

Water Vapor

Clouds

More About Precipitation

Evaporation

When puddles dry up it's called **evaporation**.

Have you ever hung your wet clothes out to dry? The water leaves your clothes and goes up into the air as water vapor. This is another example of evaporation.

Condensation

Water vapor in the air can change into water droplets or ice crystals. This is how clouds are made! This is called **condensation**.

Condensation happens all around you. Have you ever seen a foggy mirror after someone has taken a hot shower, or water dripping off of the outside of a cold glass of lemonade? These are examples of condensation.

Have you ever seen frost on a window after a cold winter night? This is another example of condensation.

The Water Cycle

By reading this book, you have learned that water vapor goes up into the sky and comes back down again as precipitation. This happens again and again in a big loop or cycle known as the **water cycle**.

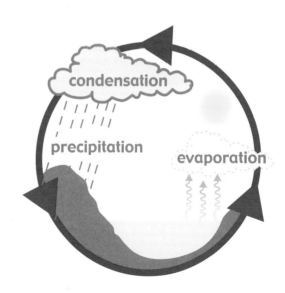

Index

About the Author
As a very young child, Stephen loved to finger paint. He had trouble learning to read in elementary school but was good at math. In high school, his favorite subject was art. In college he studied science.

Acknowledgements
Special thanks to a group of scientists at the National Center for Atmospheric Research, in Boulder Colorado, for helping us check this book for accuracy.

Photo Credits
Sincere thanks to the people at http://northfield.org, and especially to Adam Gurno (photo, page 12), Eric Hinsdale (bottom photo, page 14), and Amy Goerwitz and her daughter Jariya (right-hand photo, page 20), who provided us with their photos from the August 24, 2006, hailstorm that pummeled Northfield, Minnesota.